AF684960

Maths Sticker Workbooks
Multiplication

Wendy Clemson and David Clemson

Note to parents

This book is part of a programme of workbook titles that are designed to support school-work and make learning fun. Each page demonstrates a different mathematical concept and provides example calculations. When your child has met each concept and understood it, the practice calculations can be tried.

Many of the questions and puzzles in **Multiplication** are answered with stickers, which are found on the middle two pages of the book. Your child will need plenty of scrap paper to write down the calculations featured on each page before he or she can start to work out the answers. All answers are provided on page 16.

How to use this book:

The star sign means there is a sticker to put here on the page.

Wherever your child needs to fill in an answer, there is a blue box like this one to write in.

The calculator picture appears whenever a calculator is needed to solve a puzzle.

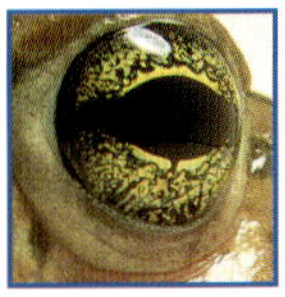

In the top left-hand corner of each page there is a space for a "reward" sticker. Your child can add it when he or she has completed the puzzles.

DORLING KINDERSLEY

LONDON • NEW YORK • MUNICH • MELBOURNE • DELHI

Multiplying numbers

Multiplication is repeated addition, adding again and again. The puzzles on these pages will refresh your multiplication knowledge. You may find the number line useful to help solve some of the puzzles.

Number line

The frog jumps along the number line in twos. Start at the beginning of the number line and move your finger along the line 2 numbers for each jump. Use the number line to solve the puzzles on the right.

0 1 2 3 4 5 6 7 8 9 10 11 12 13 14 15 16 17 18 19 20 21 22

Signs and symbols

Multiplication puzzles are usually written out using signs and symbols, as shown below. Solve the puzzles on these pages by setting out your calculations with these signs.

$3 \times 4 = 12$

This sign means multiply or times.

This sign means equals.

This number is the answer, or product.

Calculator challenge

- Key in 4 + 4 = What is the answer?
- Leave the answer on screen and press =
- Keep pressing =
- What is added to the display each time?

☐

Hungry tadpoles

A baby tadpole eats 7 insects in 1 day.

How many insects will the tadpole eat in 4 days? ☐

How many insects will the tadpole eat in 6 days? ☐

Frogspawn

A local pond is full of frogspawn. There are 6 eggs in each blob of frogspawn.

If there are 14 blobs of frogspawn in 1 pond, how many blobs will there be in 2 ponds? ☐

How many eggs will there be in 5 blobs of spawn? ☐

How many eggs will there be in 3 blobs of spawn? ☐

Leap frog in twos

Fill in the boxes to show where the frog lands in these jumps.

12 14 ☐

20 ☐ 24

32 ☐ ☐ 38

Speedy grasshopper

A grasshopper leaps in fours. How many leaps can it do in total on this number line?

☐

Circle the numbers the grasshopper can land on.

26 15 40 32

24 25 26 27 28 29 30 31 32 33 34 35 36 37 38 39 40

Growing froglet

A froglet eats 12 flies in 1 day.

How many flies will it eat in 2 days?

☐

How many flies will 1 froglet eat in 3 days?

☐

Frogs' legs

A frog has 5 digits on each limb.

How many digits has the frog got on all four limbs?

☆

Find the sticker answers to these calculations.

☆ x 5 = 10

☆ x 5 = 60

Fast eaters

This hungry toad gulps down 11 flies in 1 hour.

How many flies can the toad eat in 5 hours?

☆

Find the sticker answers to these calculations.

2 x ☆ = 22

7 x 11 = ☆

Times tables

Become a times table detective. Study the notes and see if you can find the missing numbers in the number patterns on these pages.

Moving in twos

The products of the 2 times table are all even numbers. Write in the missing numbers in the boxes below.

2 ☐ 6 8 ☐ ☐ ☐ 16 18

Grouping in fives

Write in the products of the 5 times table in the column below.

5 ☐ ☐ ☐ ☐ ☐ ☐ ☐ ☐ ☐ ☐

What pattern do the numbers make?

Does this pattern work for any series of numbers multiplied by 5?

Tick the correct answer.

YES NO

Agent 7

You are meeting a secret agent in a high-security building. Before you can enter the building, you need to find out the door code. You have been sent two notes, as shown below.

29 24 36 35 18
76 40 42 49 19 48
20 70 56 63 64

The door code is the products of the 7 times table in the number note.

35 is the first product of the 7 times table.

What is the door code?

3 5 _ _ _ _ _ _ _ _ _ _ _

You then receive a message – the second door code is the products of the 8 times table.

What is the second door code?

_ _ _ _ _ _ _ _ _ _

When you meet the agent, you need to give him another code – the number that appears in both the 7 and 8 times tables.

What is it? ☐

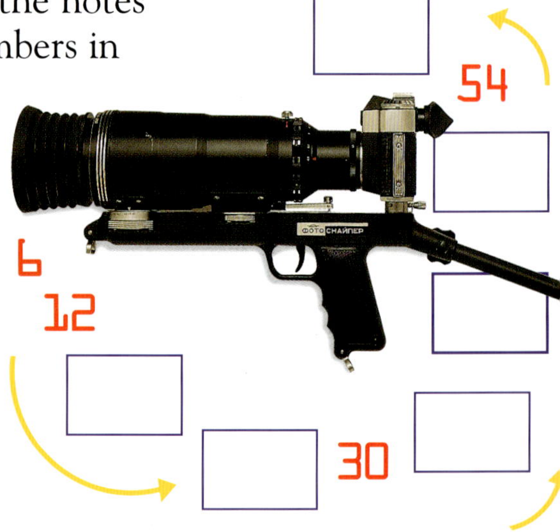

Spy camera sixes

Study the trail of the 6 times table around the spy camera above, and fill in the missing numbers.

Secret film

The key pictures on this film are arranged following the products in the 3 times table. Which numbers also appear in the 6 times table?

Write your answers in these boxes.

☐ ☐ ☐ ☐ ☐

Which numbers also appear in the 4 times table?

☐ ☐

Crack the codes

Study the one hundred square below. Circle the numbers, in different colours, to show counting in threes, fours, and twelves. Check the patterns you make against those below and put in the stickers.

1	2	3	4	5	6	7	8	9	10
11	12	13	14	15	16	17	18	19	20
21	22	23	24	25	26	27	28	29	30
31	32	33	34	35	36	37	38	39	40
41	42	43	44	45	46	47	48	49	50
51	52	53	54	55	56	57	58	59	60
61	62	63	64	65	66	67	68	69	70
71	72	73	74	75	76	77	78	79	80
81	82	83	84	85	86	87	88	89	90
91	92	93	94	95	96	97	98	99	100

Square numbers

When we multiply a number by itself, the product is a square number. Fill in the boxes and find the sticker answers to complete this square number puzzle.

1x1 2x2 3x3 4x4 5x5 6x6 7x7

8x8 9x9 10x10

11x11

12x12

Calculator challenge

9x table

- Using a calculator, multiply any number you like by 9. Add the digits of the answer together. Keep adding until only one digit remains.
- Is it **always** 9?

Tick the correct answer.

YES NO

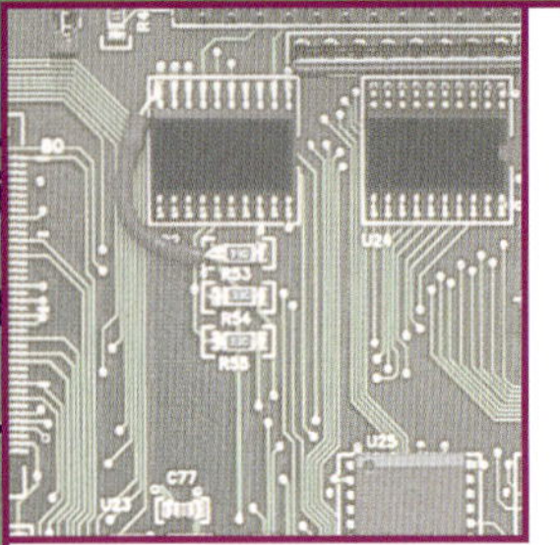

Multiplication practice

Learn more about multiplying by solving these puzzles. When you know your times tables really well, you will be able to turn calculations around, too.

Number game

There is a number game on this computer. Each number keyed in is multiplied by 7 to give a product on the screen. Write in the boxes the products that will appear.

IN		OUT
3	→	☐
12	→	☐
9	→	☐

If the computer is multiplying by 12, which numbers have been keyed in?

IN		OUT
☐	→	96
☐	→	60

Sticker calculations

Find the correct stickers to show how the computer is producing these answers.

IN				OUT
9	→	☆	→	45
11	→	☆	→	33

Mixed calculations

Sometimes addition, subtraction, and multiplication are all needed in calculations. Follow the examples below to see how this is done.

You need to study the position of the brackets carefully.

(4x3) + 2 = ?
12 + 2 = 14

4 x (3 + 2) = ?
4 x 5 = 20

Remember to do the calculation in brackets first.

Try to solve these puzzles by doing the calculations in the same order as shown in the examples.

(4 + 5) x 8 = ☐

(10 x 2) – 7 = ☐

(11 x 7) + 20 = ☐

Rows and columns

Items can be arranged in rows and columns. If you know your times tables well, you can also do this with figures.

Look at how the CDs have been arranged below and study the calculations to see how this works.

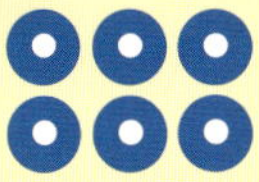

3 x 2 = 6

Columns Rows

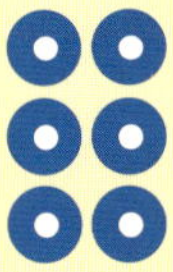

2 x 3 = 6

Columns Rows

You can multiply 2 by 3 or 3 by 2 to get the same answer.

2 x 3 = 3 x 2

Organizing files

The 16 files in this computer are represented by icons. The icons have been arranged in 4 rows and 4 columns.

If the files are arranged in 2 rows, how many columns will there be?

If the files are arranged in 16 rows, how many columns will there be?

If there were 15 files in 3 rows, how many columns would there be?

4 x 3

Rows Columns

2 x 6

Rows Columns

Virus frenzy

A computer expert has found 12 viruses on a computer. The viruses come in groups of 3. Position the stickers in rows and columns to show these 3 possible arrangements for the viruses on the computer screen.

How many more possible arrangements are there?

1 x 12

Row Columns

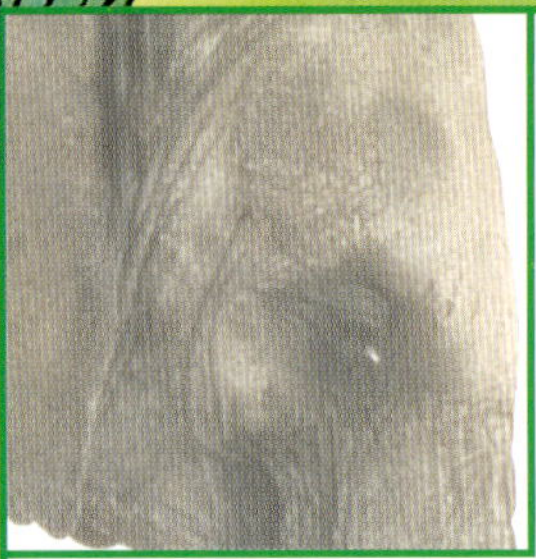

Long multiplication

Make quick work of multiplying large numbers by using long multiplication. Study the examples on this page and then try to solve the long multiplication puzzles.

Browsing elephant

If there are 32 elephants in a herd, you can use long multiplication to find out how many elephants there will be in several herds.

Set out a simple multiplication vertically to practise the first steps of long multiplication.

$32 \times 9 = ?$

	Hundreds	Tens	Units
		3	2
x			9
	2	8	8
		1	

First, multiply 9 by 2.

9 x 2 = 18. Put 8 in the units column and carry over the 1 ten.

Then, multiply 9 by 30

9 x 30 = 270. Add the 1 ten carried over and put 8 in the tens column, and 2 in the hundreds column.

The example below shows one method of long multiplication: multiplying the units first.

$32 \times 19 = ?$

	Hundreds	Tens	Units
		3	2
x		1	9
	2	8	8
		1	
	3	2	0
Find the sum of the products to get the answer.	6	0	8
	1		

First, multiply 32 by the units (9): 9 x 2, then 9 x 30.

9 x 2 = 18. Put 8 in the units column and carry over the 1 ten.

9 x 30 = 270

Add the 1 ten carried over.

Put 0 in the units column to show you are multiplying the tens.

Then, multiply 32 by the tens: 10 x 2, then 10 x 30.

You can also find the product of a long multiplication by multiplying the tens first.

		3	2
x		1	9
	3	2	0
Find the sum of the products to get the answer.	2	8	8
	6	0	8
	1		

First multiply 32 by 10.

Then, multiply 32 by 9.

Practise a long multiplication here.

64
x 32

What is the answer?

64 x 32 = ☐

Puzzle it out

Solve these long multiplication puzzles and find the correct sticker answers.

99
x 92

172
x 56

154
x 60

2/3 Multiplying numbers

Frogs' legs

Fast eaters

4/5 Times tables

Crack the codes

3	4	12

Square numbers

36	25	9
81	64	49

6/7 Multiplication practice

Sticker calculations

Virus frenzy

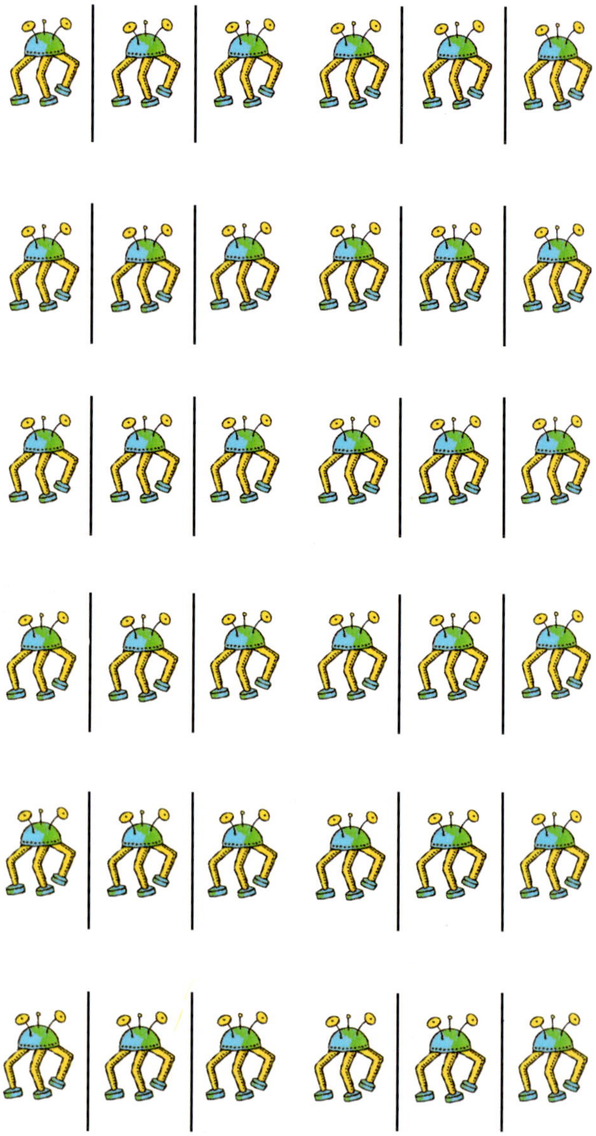

Reward stickers

When a page is completed and the answers checked, reward yourself with the right sticker.

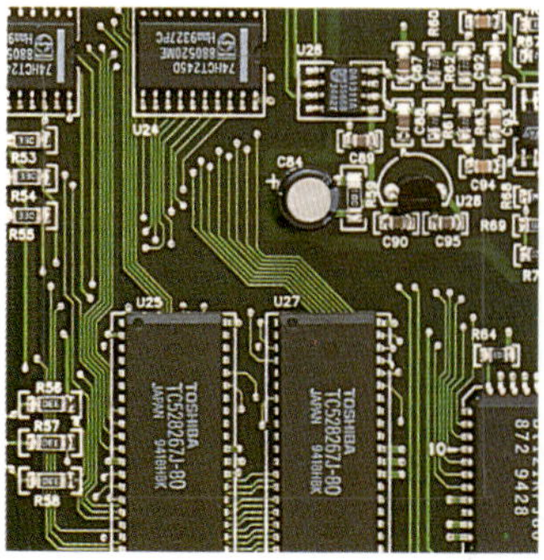

8 Long multiplication
Puzzle it out

9240

9632 9108

10/11 Decimals
Negative film sections

0.4 orange 0.6 green

0.2 green 0.8 orange

0.1 orange 0.9 green

0.3 green 0.7 orange

12/13 Area and perimeter
Area and perimeter practice

8m | 4m | $2m^2$ | 6m | $3m^2$ | 8m | $1m^2$ | $4m^2$

14/15 Scoring baskets
Multiplication stickers

4 x 3	8 x 5	9 x 2	6 x 3	6 x 5	9 x 9	7 x 5	4 x 8
8 x 2	4 x 7	8 x 9	9 x 3	6 x 6	2 x 4	7 x 6	4 x 2
8 x 8	5 x 2	8 x 7	9 x 4	7 x 2	5 x 7	3 x 5	4 x 6
7 x 7	5 x 3	5 x 5	9 x 5	9 x 7	7 x 3	2 x 5	5 x 4
8 x 3	8 x 6	6 x 2	9 x 6	9 x 8	7 x 4	4 x 4	8 x 4

Multiplying large numbers

Here are some more long multiplications for you to answer and practise your multiplying skills. You will need plenty of scrap paper to solve these calculations.

Pride of lions

If there are 18 lions in a pride or family, how many lions will there be in the number of prides shown below? Fill in the boxes with the correct answers.

☐ lions in 15 prides.

☐ lions in 57 prides.

☐ lions in 64 prides.

If there are 27 lions in 1 pride, how many will there be in 39 prides?

☐

Thirsty rhinos

If 1 water hole has enough water for 13 rhinos to drink from, how many rhinos can drink from the number of water holes shown below?

☐ rhinos can drink from 12 water holes.

☐ rhinos can drink from 23 water holes.

☐ rhinos can drink from 47 water holes.

If 1 water hole has enough water for 22 baby rhinos to drink from, how many baby rhinos can drink from 14 water holes?

☐

On safari

102 people visit the safari park each day. How many people will visit in the following number of days?

People ☐ in 14 days.

People ☐ in 17 days.

People ☐ in 2 ½ days.

Decimals

Some calculations do not give whole number answers – they may show parts of a whole number as a decimal or fraction.

Calculator decimals

A calculator shows parts of a whole number as a decimal. Any number appearing to the right of the decimal point is a decimal.

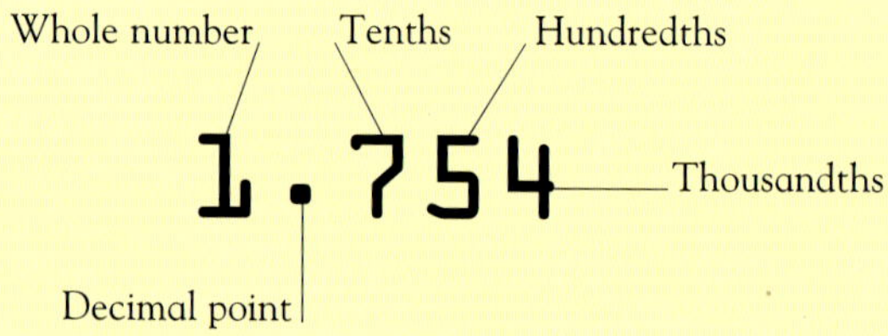

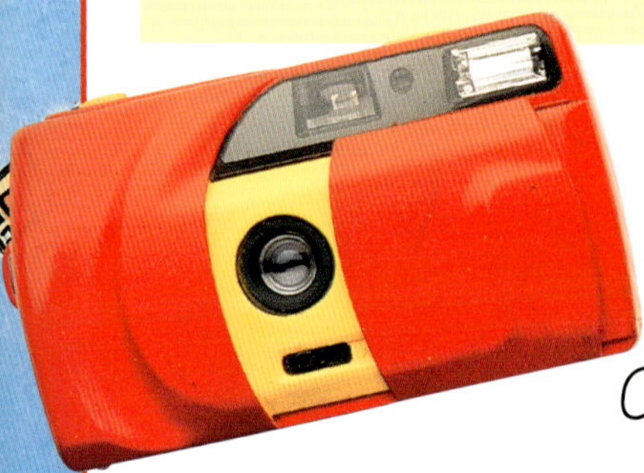

Calculator practice

Use a calculator to solve the puzzles below. Can you see what is happening to the decimal point in these calculations?

0.1 x 10 = ☐ 0.01 x 10 = ☐ 1.0 x 10 = ☐

Enlarging photographs

These reduced photographs can be enlarged in an enlarging machine. Solve the calculations below to find out the size of each enlarged photograph.

IN

0.2

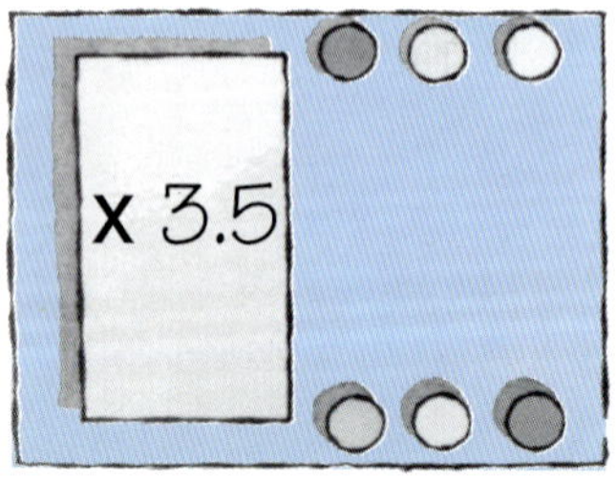

OUT ☐

IN

0.8

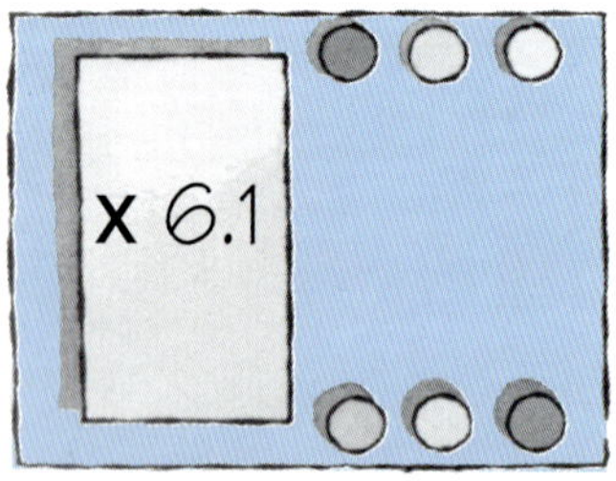

OUT ☐

IN

0.3

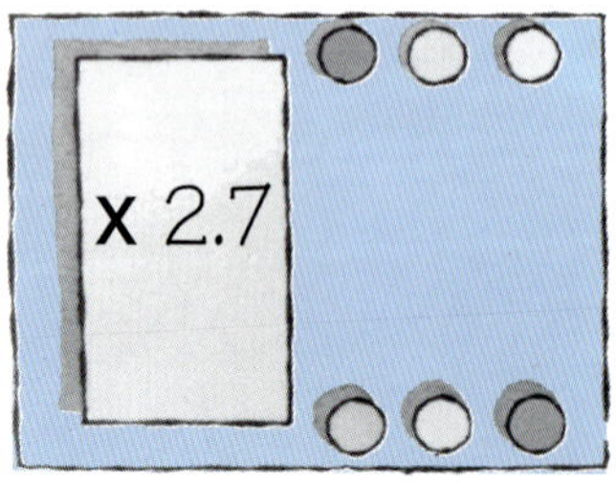

OUT ☐

Puzzle it out

Solve these calculations and study what happens to the decimal points.

1.2 x 3.4 = ☐

11.01 x 5.00 = ☐

11.01 x 50.0 = ☐

2.5 x 2.5 = ☐

0.8 x 0.09 = ☐

Decimal fractions

The negative film below can be divided into sections in the same way as a whole number can be divided into fractions or decimals. Each of the 10 sections represent one-tenth or 0.1 of the whole film.

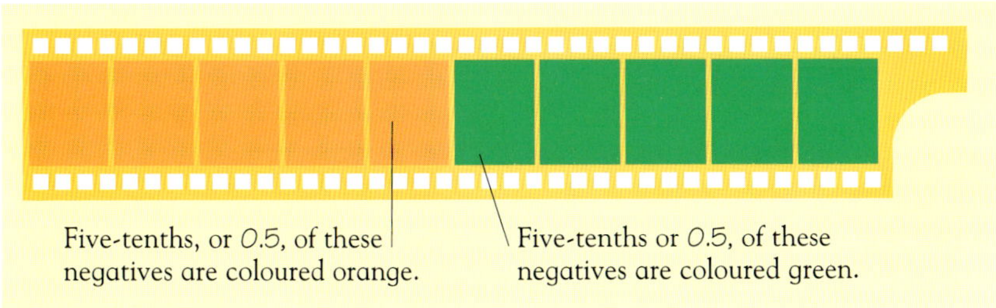

Choose the correct sticker answers to show the decimal fractions of the coloured film sections below.

Photograph sizes

The photographs below have been reduced to 0.75, 0.5, and 0.25 of the full-size photograph. Which is which?

Full-size photograph

Half-size reduction

Quarter-size reduction

Three-quarter-size reduction

Area and perimeter

Learn about area and perimeter by solving these castle puzzles. The area is found here by adding up all the units in a floor space, and the perimeter by adding up the lengths of the sides of the floor space.

Calculating area

The area of the castle tower can be found by counting the number of units or squares in the floor plan. The area is 9 squares.

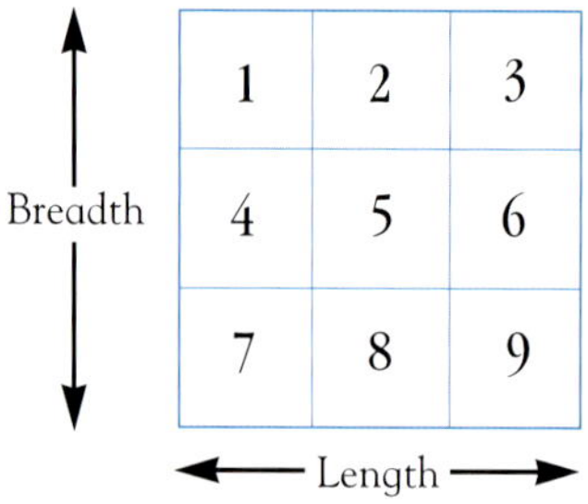

If the sides of each square represent 1 metre, the area measures 9 square metres or 9 m^2 for short.

You can also find the area of any square or rectangle by doing a multiplication calculation, as shown below.

> Length x Breadth = Area
> The castle tower area is:
> 3 m x 3 m = 9 m^2

Solve these puzzles to practise finding area.

Length		Breadth		Area	
3 m	x	4 m	=	☐	m^2
3 m	x	2 m	=	☐	m^2

Castle floor plan

Study the castle floor plan below and the key to the right. Using your knowledge of area and perimeter solve the puzzles opposite.

Castle floor plan key

- Armoury
- Dungeon
- Kitchen
- Great Hall
- Haunted corridor

Calculating perimeter

The perimeter of the castle gatehouse can be found by adding up the outside edges of the squares on the floor plan. The perimeter is 16 sides of squares.

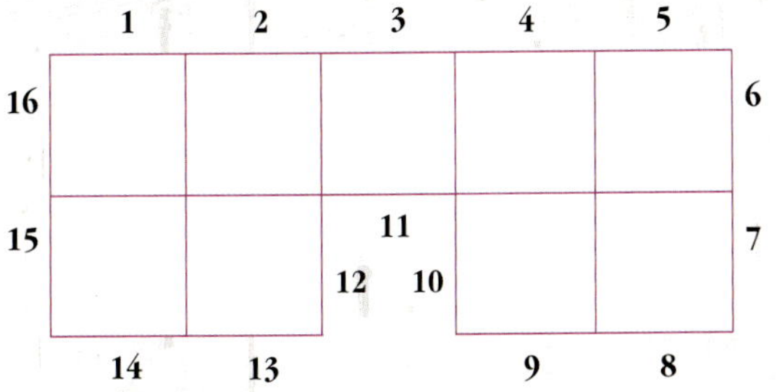

If the sides of each square represents 1 metre, the perimeter measures 16 metres or 16 m for short.

Area and perimeter practice

Find the correct sticker answers for the puzzles below to show the area and perimeter of these diagrams. The sides of the squares in each shape represent 1 m.

	Area	Perimeter

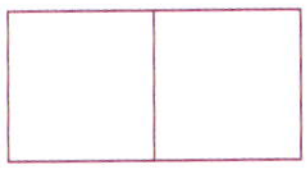		
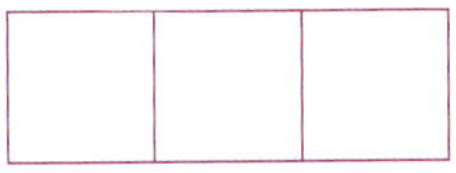		
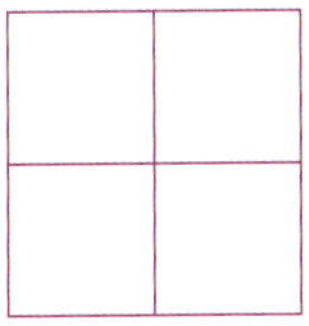		

Castle area

What are the areas of the rooms of the castle shown below? Use the floor plan to help you work out the calculations.

Armoury		m^2
Kitchen		m^2
Dungeon		m^2
Great Hall		m^2
Haunted corridor		m^2

What do you notice about these areas?

Castle perimeter

Now, work out the perimeters for all the castle rooms.

Armoury		m
Kitchen		m
Dungeon		m
Great Hall		m
Haunted corridor		m

Are the perimeters all the same? Tick the correct box.

Yes No

Scoring baskets

You will need

- Two players • Thin card
- Scissors • 40 multiplication stickers from the sticker sheet
- Scrap paper and pencils to work out the answers

Making the cards

Stick the multiplication stickers on to the thin card. Then, cut out a pack of 40 equal-sized multiplication cards.

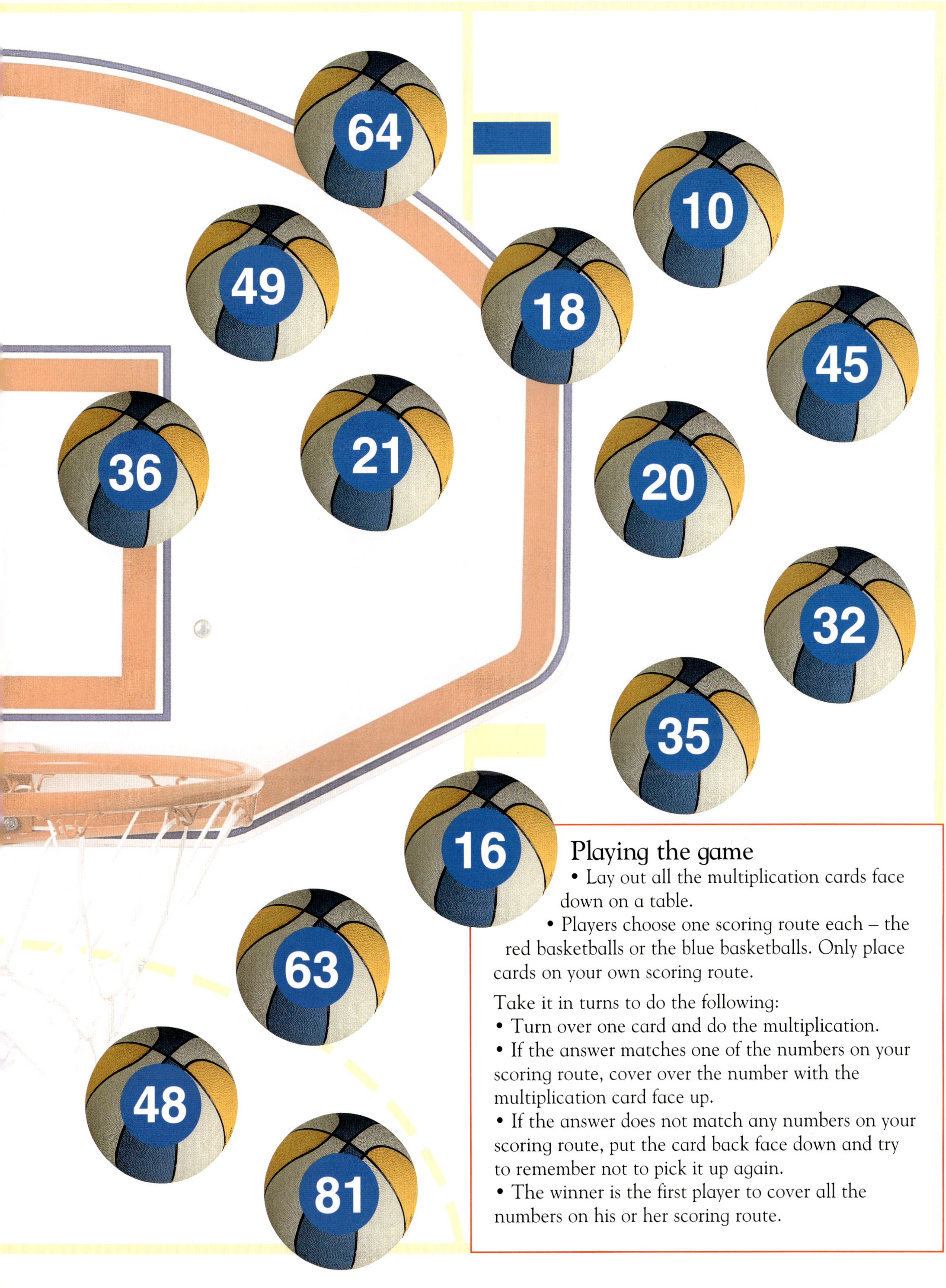

Playing the game

- Lay out all the multiplication cards face down on a table.
- Players choose one scoring route each – the red basketballs or the blue basketballs. Only place cards on your own scoring route.

Take it in turns to do the following:

- Turn over one card and do the multiplication.
- If the answer matches one of the numbers on your scoring route, cover over the number with the multiplication card face up.
- If the answer does not match any numbers on your scoring route, put the card back face down and try to remember not to pick it up again.
- The winner is the first player to cover all the numbers on his or her scoring route.

Answers

Pages 2/3 Multiplying numbers

Leap frog
The frog lands in the following places:
12 14 **16**
20 **22** 24
32 **34 36** 38

Speedy grasshopper
The grasshopper can do **10** leaps on the number line: **10** x 4 = 40
The grasshopper can land on these numbers: **40** and **32**

Calculator challenge
4 is added to the display each time.

Hungry tadpoles
In 4 days the tadpole eats **28** insects:
7 x 4 = **28**
In 6 days the tadpole eats **42** insects:
7 x 6 = **42**

Frogspawn
There will be **30** eggs in 5 blobs of frogspawn:
6 x 5 = **30**
There will be **18** eggs in 3 blobs of frogspawn:
6 x 3 = **18**
There will be **28** blobs of frogspawn in 2 ponds:
14 x 2 = **28**

Growing froglet
The froglet will eat **24** flies in 2 days:
12 x 2 = **24**
The froglet will eat **36** flies in 3 days:
12 x 3 = **36**

Frogs' legs
The frog has **20** digits in total:
5 x 4 = **20**
2 x 5 = 10   **12** x 5 = 60

Fast eaters
The toad will eat **55** flies in 5 hours:
11 x 5 = **55**
2 x 11 = 22   7 x 11 = **77**

Pages 4/5 Times tables

Moving in twos
2 4 6 8 10 12 14 16 18

Grouping in fives
The number pattern is 5 0
The answer is **YES**, the pattern works for any number multiplied by 5.

Agent 7
The door code is 3 5 4 2 4 9 7 0 5 6 6 3
The second door code is 2 4 4 0 4 8 5 6 6 4
The code to give to the agent is 56.

Spy camera sixes
6 12 18 24 30 36 42 48 54 60

Secret film
The following numbers appear in both the 3 and 6 times tables:
6 12 18 24 30
The following numbers also appear in the 4 times table: 12 24

Crack the codes
From top to bottom, the patterns show counting in threes, twelves, and fours.

Square numbers
1 x 1 = 1   6 x 6 = 36   11 x 11 = 121
2 x 2 = 4   7 x 7 = 49   12 x 12 = 144
3 x 3 = 9   8 x 8 = 64
4 x 4 = 16   9 x 9 = 81
5 x 5 = 25   10 x 10 = 100

Calculator challenge
The answer is **YES.**

Pages 6/7 Multiplication practice

Number game
3 x 7 = 21   8 x 12 = 96
12 x 7 = 84   5 x 12 = 60
9 x 7 = 63

Sticker calculations
9 x 5 = 45   11 x 3 = 33

Mixed calculations
(4 + 5) x 8 = 72   (10 x 2) – 7 = 13
(11 x 7) + 20 = 97

Organizing files
If the files are arranged in 2 rows there will be 8 columns: 2 x 8 = 16.
If the files are arranged in 16 rows there will be 1 column: 16 x 1 = 16.
If there are 15 files in 3 rows, there will be 5 columns: 5 x 3 = 15.

Virus frenzy
The 3 possible sticker combinations are:

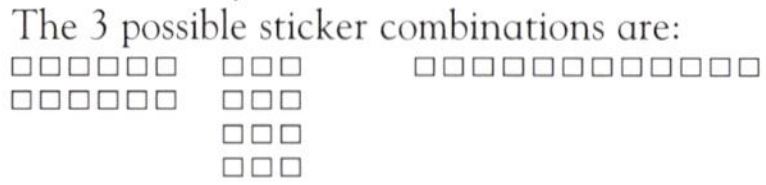

There are 3 more possible arrangements:

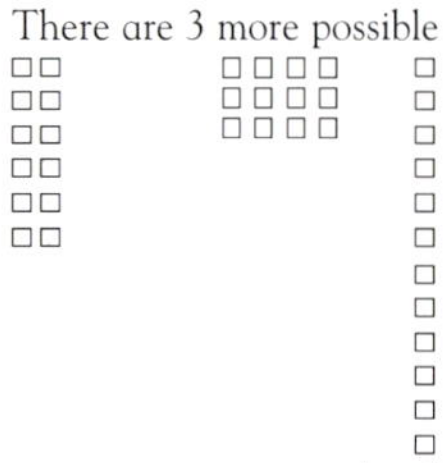

Page 8 Long multiplication

Browsing elephant
64 x 32 = **2048**

Puzzle it out
99 x 92 = **9108**   172 x 56 = **9632**
154 x 60 = **9240**

Page 9 Multiplying long numbers

Pride of lions
There will be **270** lions in 15 prides:
18 x 15 = **270**
There will be **1026** lions in 57 prides:
18 x 57 = **1026**
There will be **1152** lions in 64 prides:
18 x 64 = **1152**
If there are 27 lions in 1 pride, there will be **1053** lions in 39 prides: 27 x 39 = **1053**

Thirsty rhinos
156 rhinos can drink from 12 water holes:
13 x 12 = **156**
299 rhinos can drink from 23 water holes:
13 x 23 = **299**
611 rhinos can drink from 47 water holes:
13 x 47 = **611**
If 22 baby rhinos can drink from 1 water hole, **308** baby rhinos can drink from 14 water holes:
22 x 14 = **308**

On safari
1428 people will visit the park in 14 days:
102 x 14 = **1428**
1734 people will visit the park in 17 days:
102 x 17 = **1734**
255 people will visit the park in $2\frac{1}{2}$ days:
102 x $2\frac{1}{2}$ = **255**

Pages 10/11 Decimals

Calculator practice
0.1 x 10 = **1**   0.01 x 10 = **0.1**
1.0 x 10 = **10**

Enlarging photographs
0.2 x 3.5 = **0.7**   0.8 x 6.1 = **4.88**
0.3 x 2.7 = **0.81**

Puzzle it out
1.2 x 3.4 = **4.08**
11.01 x 5.00 = **55.05**
11.01 x 50.0 = **550.5**
2.5 x 2.5 = **6.25**
0.8 x 0.09 = **0.072**

Decimal fractions
The sections, from top to bottom, are as follows:
0.20 green and **0.80** orange
0.40 orange and **0.60** green
0.30 green and **0.70** orange
0.10 orange and **0.90** green

Photograph sizes
The sizes, from top to bottom, are as follows:
0.50, 0.25, 0.75

Pages 12/13 Area and perimeter

Tower area
3 x 4 = **12 m²**   3 x 2 = **6 m²**

Area and perimeter practice
From top to bottom:

Area	Perimeter
1 m²	**4 m**
2 m²	**6 m**
3 m²	**8 m**
4 m²	**8 m**

	Castle area	Castle perimeter
Armoury	**9 m²**	**16 m**
Kitchen	**9 m²**	**20 m**
Dungeon	**9 m²**	**16 m**
Great Hall	**9 m²**	**12 m**
Haunted corridor	**9 m²**	**20 m**

The areas are all the same.
The perimeters are NOT all the same.

A DORLING KINDERSLEY BOOK

Project Editor Patricia Grogan
Designers Kate Eagar and Caroline Potts
Assistant Designer Jacqueline Gooden
Managing Editor Jane Yorke
Managing Art Editor Chris Scollen
Jacket Designer Mark Haygarth
Production Ruth Cobb
Illustrations Sally Kindberg

Special photography on pages 8–9 by Philip Dowell
Photography by Dave King, Susanna Price, Matthew Ward

Picture credits: H. Keith Melton, Clandestine Services Museum: 5 BR, BC.

First published in Great Britain in 1997 by Dorling Kindersley Ltd, 80 Strand, London WC2R ORL

Copyright © 1997 Dorling Kindersley Limited, London

Visit us on the World Wide Web at http://www.dk.com

All rights reserved. No part of this publication may be reproduced, stored in a retrieval system, or transmitted in any form, or by any means electronic, mechanical, photocopying, recording, or otherwise, without the prior written permission of the copyright owner.

A CIP catalogue record for this book is available from the British Library

ISBN 0-7513-5676-X

Colour reproduction by Colourscan, Singapore.
Printed and bound by LREX in China